The DNA of Altruism

Win or Lose

Paul McCullough

Book: DNA of Altruism

Subtitle: Win or lose?

The total or partial reproduction of this work is forbidden without direct authorization from the author.

McCullough, Paul

Issue 1

ISBN: 9798873273058

The DNA of Altruism / Paul McCullough /

Barcelona – Spain 2023

canva.com - Pro – registered. / Texts by MSWord.

I want to dedicate this book to my mother Cleide.

She has always lived with all her senses placed where the need exists and has existed.

My eternal example of heart and selflessness.

Index

Foreword

By Ana Rachid – Writer, Consultant, Mentor and Founder of Innovare Per Crescere.

When I received Paul's invitation to write the foreword to this book, at first, I felt a great honour and an immense emotion, and then when the adrenaline subsided, I thought what a great responsibility.

And when I learned that the theme is Altruism, something else touched me deeply, because it is part of my daily life and essence. At the same moment that Paul was talking about the book, the simplest definition of Altruism automatically came up, which is the person who gives himself to others without expecting anything in return, and that it is a noble feeling.

I confess that I was inquisitive to read the book and know the nuances and approaches developed.

While reading the book, I connected with every chapter and it was like "diving deep" into each concept, story, and how altruism, if understood well, can influence a person's life so much. And that can sometimes be a trap in today's world, talking about Altruism is good, but being altruistic is more difficult, even more so when we experience violence in the most varied forms, and man has a negative emotional charge and transmits it to people and environments. Here, of course, there is an incongruity between speaking and doing.

My feeling today is that evaluating the current culture, the values, and the social standards discourages us from having altruistic behaviours.

In this way, the altruism that we know in the simplest definition that I mentioned, is still incipient in us, and many times, we distort what is altruism in attitudes and behaviours that are harmful to us and other people, and that will lead us to mental illnesses, which we follow so much in today's world, and which require a lot of time and effort to change.

It is in this context that Paul, brilliantly, involves us in the history of human evolution that for a long time, man has had attitudes of care and preservation with people close to him, and that altruism it is inherent in man.

As well, our genetics influence the body and mind, the environment and the environment's interaction with people, contributing to the changes we have been experiencing for thousands of years.

This book, therefore, is an invitation for you to explore all the engaging chapters and delve into the Psychopathologies of Altruism that brighten the types.

I invite you to identify with your difficulties because I identified with several concepts, that made me change my view of what Altruism is.

In the chapter True Altruism, Paul leads us to understand brilliantly and simply, how this noble feeling can be developed in us, through self-knowledge and where we can contribute positively to ourselves, to people, to the environment and the world.

It is a positive energy generated by us that will contribute to prosperity.

The path of Altruism requires balance and common sense, and I quote what Paul mentions:
"That development does not despise the foundations, just as a house is not just the foundation; there is no house without a foundation."

Yes, the path to change is sometimes more difficult, but from the moment we choose to be better, new paths and opportunities will be present to contribute to our evolution. And by walking more firmly towards altruism, we will hardly return to the previous stages, because of our strength and firmness.

This book is the door for you to step in and transform your personal development, open new doors and frontiers and boost your Altruism.

Happy reading!

Purpose of the Book

This book arose from a conversation with mentor and writer Sidnei de Oliveira. We discussed how, throughout our lives, we encounter situations in which we feel compelled to help others.

I believe that, at some point in life, we all feel challenged to offer help, whether altruistically or charitably, overcoming individual needs or situations involving countless people. How many times have I needed some help myself?

Our conversation has evolved to the point of raising questions about the extent to which we are truly altruistic. We spent some time discussing the limits of this attitude, how it affects ourselves, and how it affects others, especially those in need of assistance.

During my meetings with Sidnei, it's common for topics like this to be revisited, even when we're in an informal conversation.

At one point, Sidnei challenged me: "Paul, write a book on this topic. I'd like to read your perspective."

I accepted the challenge, and here it is. This book is a little reflection, and you are invited to reflect with us.

Paul.

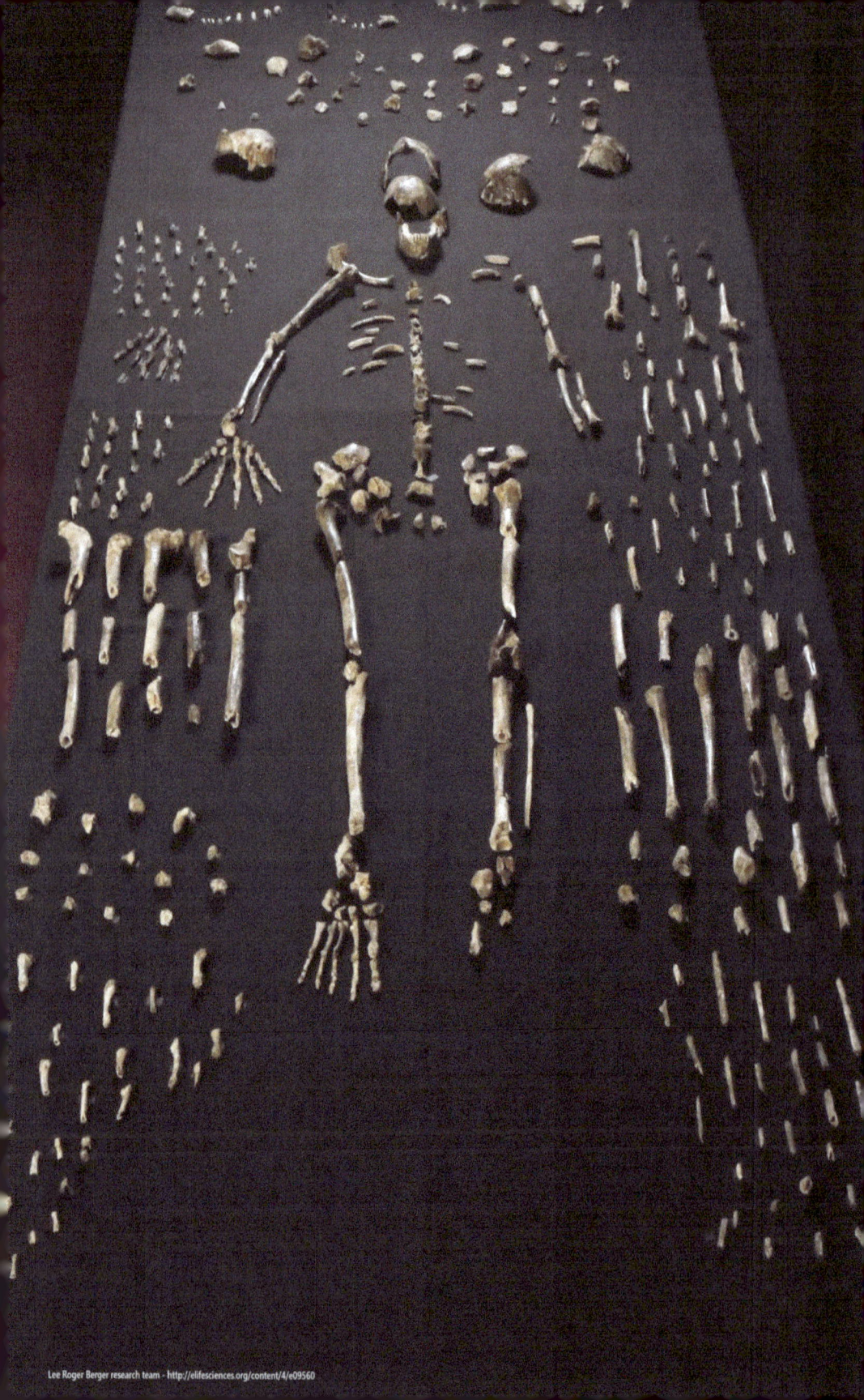

Chapter 1

The Archaeology of Altruism

"How old are you?

Remember the time when you were very young? Those TV ads? What about the clothes when you were about 12 or 13? What was fashion like back then? What about cars, remember?

If you had to choose between buying a new car and a much older one, how would you feel about buying the old one?

Of course, these questions have many possible answers. It was just a provocation for you to look at the differences of the past, to see how things have changed. It's not about liking the past or the present anymore, but about that antique face. The culture of the past is not the same as it is today.

The time that passed was not that much. Time in our society is interesting, everything seems to be changing, including our customs which change almost from one year to the next. Our behaviour changes as the more general culture spreads when new slang

emerges in people's language, often coming from what we see on TV or social media.

Recently, my 6-year-old son sat down next to me and started counting numbers. It went up to 100. For him, it was incredible since it's not that simple for a 6-year-old to count to 100. So, the challenge increased, we counted to 200, and he wanted to get to 1 million.

I explained that to get to 1 million, we would have to do the same exercise, count to a thousand, and do it a thousand times.

Do you know how boring it is to count to 1000? Imagine having to tell it a thousand times.

I used the example of the year 2023 it's 365 days counted 2023 times.

My son loves numbers and was impressed.

I wondered... How many things have changed from 2023 years ago? The fashion, the way of thinking, etc.

I told my son that the human being appeared about 300,000 years ago. He was amazed, but he had no idea of the proportion, imagine, it's hard for us adults, anyway, he was impressed.

Looking back at that time is fascinating. Without going into detail about biological development, I would like to talk a little about the topic, talking about it here will be a little different from the explanation I gave him because it was much more basic. My little one asked which man was the oldest if he had lived a million years ago.

How do you get a six-year-old boy to understand these things?

I told him that we ("we" means people like Dad) use the term "Homo" to refer to a set of hominid species belonging to the family of hominids.

They were primates, which includes contemporary humans. Of course, I didn't say it like that, I'm talking to you now, but basically, that's what I tried to explain.

These species share distinguishing features such as bipedalism (the ability to walk on two legs), a wider and rounder skull compared to other primates, and an increase in brain size.

The genus Homo emerged about 2 to 2.5 million years ago, with the first species. One of the oldest known is Homo habilis, which inhabited East Africa. This species is often regarded as a predecessor of humans, exhibiting several more advanced characteristics than earlier species.

Subsequently, Homo erectus emerged around 2 million years ago, becoming known as the first hominid to disperse outside of Africa, colonizing regions as far away as Asia and Europe. This species has shown remarkable advances in the manufacture of tools and the adaptation to different environments.

Homo Neanderthals, commonly called Neanderthals, represent another species of the genus Homo that coexisted with Homo sapiens. Neanderthals inhabited Europe and parts of Western Asia for a considerable period.

Finally, Homo sapiens, i.e. modern humans, emerged in Africa approximately 300,000 years ago, eventually expanding beyond the African continent replacing or interacting with other species of the genus Homo.

Surprisingly, I found that talking like this to my son has a calming effect, he slept soundly.

I hope you have a little more patience than my son and don't sleep, because I want to talk about something more recent than 2 and a half million years.

Remember, though, that just a few years in your life can mean big changes. Great, don't lose that perspective at any point in this little book. I say this because over time the factors that led to evolution have changed our perception of things, and despite the complexity that we have arrived at.

At present, the essence of the reason for the behaviour remains the same. Yes, the theme of the book remains Altruism. Understanding this human trait will require you to understand human development. Knowing this will make you look at this topic with different eyes (open eyes). Later, when I talk about the psychology or even the anthropology of altruism, you should take into consideration this evolution and the issues raised here in this chapter.

I begin by introducing the history of beings that populated part of southern Africa 356,000 years ago.

Homo Naledi.[1]

The name "Homo naledi" was given to a hominid species discovered in the Dinaledi chamber in South Africa in 2013.

The term "naledi" comes from Sesotho, a local language, and means "star." The choice of name

---

[1] See at the end in references.

reflects the location of the discovery, the Rising Star Cave, whose name in Sesotho is "Star Cave".

The term "Homo" indicates inclusion in the broader category of hominids, which also encompasses modern humans.

"Homo" refers to the human genus and includes the species considered "human" or called "man."

These species are part of the hominid primates of the tribe Hominin. They are characterized by walking on two legs, having flat feet with aligned non-grasping toes, an upright skull, and a larger brain compared to

other primates. The genus Homo encompasses modern human species, such as Homo sapiens, and their closest relatives.

The Naledi were bipedal and had very wide arms, which indicates partial life in the trees, but they could

walk upright and ate cereals and plants, but there is still no evidence that they could even eat meat.

They are known to have mastered fire and there is some evidence that they used stone tools to possibly harvest grain and fruit. Although they were Homo, they were not part of the direct line of evolution "sapiens", they were more primitive beings.

One of the most intriguing and fantastic things about Homo naledi was the discovery of a place where they buried their dead.

I don't know how much it impacts you, but Homo naledi 356,000 years ago cared about burying their dead.

And why is this so important? Remember, we're talking about the formation of human behaviour, even more, think about variables, they didn't form part of the formation of the genus sapiens. What a fantastic question, what causes this behaviour to occur?

When a Naledi member died, there was sadness. Nowadays we see this in modern primates and some mammals, such as elephants, for example, or gorillas and chimpanzees. However, burying is something that goes further.

In the comment of Agustin Fuentes, Evolutionary Anthropologist (Netflix Film - Unknown - Cave of Bones, 2023) "not seeing a member of your family get eaten by other predators, or even not deteriorating on

the ground, for example, when falling from a tree, is an act of conscious emotion." For that, you must have some purpose, whether it is to continue with the loved one, which can be interpreted as something "spiritual" or even a feeling of preservation.

What fascinates me is to imagine that for these primitive beings 356,000 years ago, it was necessary to have a good reason for caring for someone of your kind who had died is a telling attitude that something else is at stake.

356,000 years ago, brains were far more primitive than Homo of the Sapiens lineage doing something for someone other than themselves. This is the birthplace of attitudes.

But how do we know this? Well, we know this because of some archaeological evidence, based on observations of the burial site in Rising Star Cave. The deliberate disposition of the bodies suggests some level of ritualistic behaviour. The direct connection to specific emotions is a speculation, of course, there is no way not to speculate. The Naledi transported their dead to a cave and deposited the dead in urns dug in the ground away from the entrance, for they had to climb a steep inner wall and reach a plateau and later,

from the deepest and highest part of the plateau, descend through a kind of moat until they reached the lower part where there was a natural room and there. After a whole process of hard and complex journey, they placed their dead in these urns. Why not just throw the bodies into the cave, like any primitive

animal that wanted to hide its dead? They had to walk a lot, climb a lot, and risk their lives inside that place, they had to take water and food, take material to make fire and to illuminate, on the side, to take the corpse itself, the whole process to the burial place.

Why make an urn where they could only deposit the bodies? Why did they deposit stone tools with some of them? Why make scratched prints on the walls? What's the need to record ideas?

One might say, but perhaps they did it because they had a belief that depositing bodies like this only benefited them, i.e., a personal ritual. But is this what we see in nature? Is this what we saw in the ascendancy of the sapiens lineage? Of course not. These movements always signified the preservation of the other being after this state of perpetual starvation, when life was no longer in the body, the being, that being was to be preserved, was to continue to be part of the whole and of all.

How could these brains from thousands of years before what we know today as "thought development" formulate such a thing? How could they, not being of the same lineage, "sapiens" think as such?

What made it so?

Yes, a sea of scientific speculations, however, everything points to one place, to think about the other is to preserve the genetic lineage, the continuity of the gene is greater than the existence itself. Existing, in whatever context, is what gives reason to

living together, to want, to care, to preserve and maintain and to conserve life even if it ends. Is the gene selfish or altruistic?

To be or not to be, that is the question.

Have you ever heard of Charles Darwin and his idea of "survival of the fittest"? Well, another important name to mention in this book is that of William Hamilton[2], a scientist, who added something interesting to this idea. He talked about how nature can favour traits that help not only the individual but also their close relatives. He called this "kin selection."

Hamilton also created something called "inclusive fitness." This is a way to measure an animal's success

---

[2] William Donald Hamilton FRS (1 August 1936 – 7 March 2000) was, recognized as one of the most significant evolutionary theorists of the twentieth century. Hamilton became known for his theoretical work exposing a rigorous genetic basis for the existence that was a key part of the development of the gene-centred view of evolution. It is considered one of the forerunners of sociobiology. Hamilton published important works on the sexual reasons and the evolution of sex. From 1984 until he died in 2000, he was a Royal Society Research Professor at the University of Oxford. (https://en.wikipedia.org/wiki/W.D. Hamilton)

at reproduction, considering not only its offspring but also how their actions affect relatives.

The interesting thing about this line of study is that this occurs in the animal kingdom frequently and is a known pattern.

Altruism relates to a reason that goes beyond just helping but benefiting a genetic interest. We don't perceive this line of behaviour occurring because we look at behaviour as something individual, but if we look at it from a more analytical perspective, we will find this natural occurrence. Hamilton studied many social insects, such as ants and bees. He found that some of these insects give up having their own young to help relatives be more successful in reproducing. He explained this using the idea of kin selection. It has also contributed to understanding how animals choose their mates (called Sexual Selection) and how certain traits evolve to become more attractive.

He even formulated a mathematical rule, called "Hamilton's rule," which says that altruistic behaviours (helping others without getting anything in return) evolve when the benefit to the person being helped (considering kinship) is greater than the cost to the one who helps.

(W.D. Hamilton, Evolution of Social Behaviour." It is a collection of papers by W. D. Hamilton)

Perhaps this explains the issue of Homo naledi, when the interest in somehow keeping a member of his family connected to the meaning of life, and in some way, acting towards the closest, that is, preserving. Like the dog that tries to bite its tail, the fact that someone gives meaning to another, someone, suggests the possibility of ritualistic behaviours or a rudimentary form of death consciousness. So, we have the signs of the formation of self-consciousness.

356,000 years ago, the understanding of death in human evolution, this is undoubtedly a strong indication of the first attitudes of what we understand as altruism. When a person helps their family members survive and have children, part of their genetic traits are passed on to the next generations. This is called "inclusive fitness," where the reproductive success of the family is considered.

In addition, group selection is an explanation for altruistic behaviour. If groups of people collaborate, the group may have an advantage over other groups. In this way, altruistic behaviours within the group can be favoured by natural evolution.

In addition to genetic evolution, altruistic behaviour can also be influenced by cultural evolution. As here I am talking about groups that participate in the formation of what we know today as Homo sapiens, it's important to understand that we're also talking about social norms and values that promote cooperation and altruism, it's the behaviour of the group, that can establish the genesis of what can be transmitted culturally from one generation to another.

So, what can we conclude so far? Altruism is an inseparable part of us as social and evolved beings.

It would not be fair to talk about genetics without understanding precisely how the manifestation of behaviour occurs through the mechanics of the gene. Gene expression is changes in the gene that project to the outside world, or the environment. In the same way, from the outside world, where genetic behaviour expresses itself, interacts, acquires structure, and evolves, this outer result promotes changes in structure genetics, thus contributing to ongoing genetic changes. Yes, it is a circle, which changes its shape, and this dynamic constitutes the area of science known as epigenetics.

I'll try to explain how this happens more simply: our genes determine our body and mind.

They work, and these changes within us are reflected outside, in the environment around us, and interactions with other people.

These exchanges of changes and behaviours caused by each person's genes create adaptive relationships.

For example, if I eat red fruit and get sick, I'm likely to avoid all berries in the future. However, if we meet someone who eats the same berries and has no problems, both they and I will try to understand why we react differently to the berries.

In this process, many variables can arise. For example, I might think that anyone who gets sick when eating berries has some problem or something evil. So, I teach my family to avoid everyone who gets sick with these fruits, there will be no weddings and there will be no cooperation. This means that, over time, the gene expression that will prevail will be that of those who are tolerant to berries. The bigots will weaken and diminish, perhaps disappear.

How does this happen? First, there needs to be an expression in the gene, present in both the intolerant and the tolerant person. Next, this expression needs to manifest externally in the environment, and this external interaction will influence gene expressions over time.

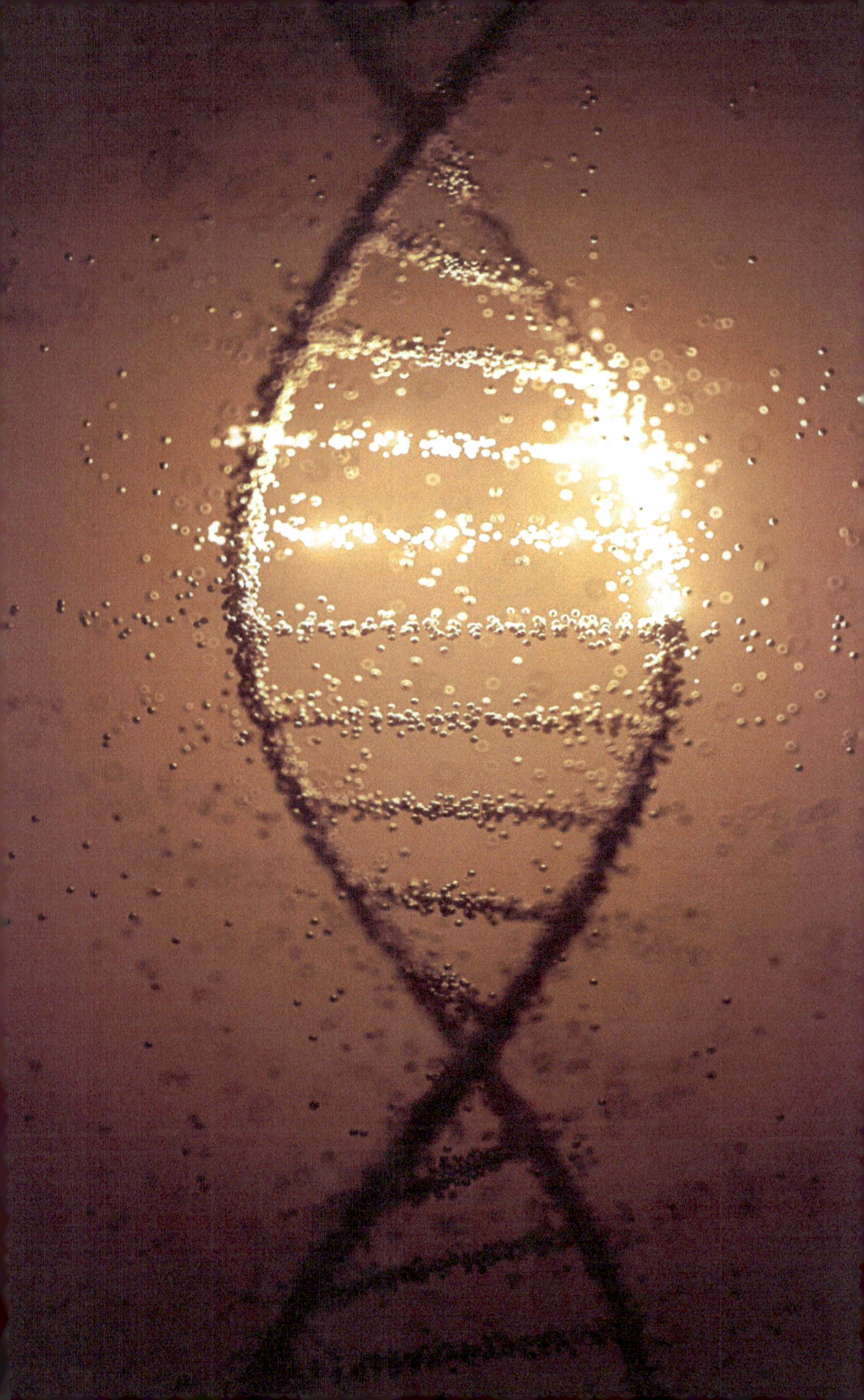

Chapter 2

The Epigenetics of Altruism

Epigenetics is a serious science, difficult in the sense that it deals with a lot of information and knowledge. In addition, it is extremely complex and constantly evolving.

The following example in no way sums up epigenetics; It's just one example of one of the tools used to understand evolutionary processes. We should not generalize the whole from a part.

A study published in the journal Nature Neuroscience in 2004 (Kim, M., Costello, J. DNA methylation: an epigenetic mark of cellular memory. Exp Mol Med 49, e322 (2017). https://doi.org/10.1038/emm.2017.10)

In studies with mice, it has been observed that the nurturing way mothers interact can influence DNA behaviour.

The mice that received the most maternal affection showed alterations in specific parts of the DNA, a process called DNA methylation, related to the stress response in the pups.

DNA methylation involves adding small chemical groups to the DNA molecule, without changing the sequence of its parts. This is crucial as it regulates the activation or deactivation of genes.

When considering environmental factors, it is possible that behaviours beneficial to life could be genetically transmitted through this DNA methylation process. The environment exerts an influence on DNA, as exemplified by the impact of chemicals on animals exposed to constant dangers.

Similarly, positive factors, such as a lack of stress and favourable eating conditions, promote healthy interactions, balance, and resources for group survival. These concepts are also related to ideas such as inclusive fitness, and sexual selection, among others, which contribute positively to the development of the species.

In epigenetics, it is relevant to highlight that these processes do not occur only in specific situations or isolated life cycles, but over longer periods, repeating themselves continuously.

When applied to human beings, we're talking about thousands of years of evolution.

The complexity of the interaction between genetics and environment is evident in the continuity of these patterns over time. The link between genetic material and environmental influences shapes not only physical traits but also behaviours passed down from generation to generation. It's like an intricate dance between inherited genetic information and the adaptations needed to thrive in specific environments.

Understanding DNA methylation and other epigenetic processes is key to unravelling the complexity that unites genetic inheritance and adaptive responses to the environment.

By analysing this phenomenon over millennia, we can see how life, in its various forms, continues and adapts its mechanisms to ensure survival and reproduction in ever-changing environments.

Remember my first proposition, how many times and how do all things change in a year?

So, let's flip the switch on the atonement of time, from the thousands of years when these interactions began to emerge to the present.

Imagine, over 356,000 years, the interactions resulting from natural motives in the constitution of

every being. These interactions shaped what we now call social relationships.

We should not underestimate the time that has elapsed, as the genetic influence on how we perceive the other, act towards the other, and adopt altruistic behaviours persists. The question that arises is whether this genetic interaction has the same impact as it did in the past, or how this pattern evolved in our development as thinking beings.

Currently, our planet is a chaotic place, especially when it comes to interpreting what is good, bad, positive, and negative to society and human behaviour. However, we still maintain some standards that help us analyse this development more critically. Firstly, in a relatively short period, society has undergone significant changes. Modern society, at most, has 2,000 to 3,000 years of evolution, compared to the previous 300,000 years, in which the structures of modern society were not present.

For example, sunsets still influence our bedtime because, over thousands of years, we've relied on sunlight to activate processes of melatonin that induce sleep. Electric light is only about 140 years old.

From an evolutionary point of view, 140 years is a drop in the ocean, and it will take much longer for electric light to have a significant impact on our circadian system.

Although electric light can interrupt our sleep today, genetic adjustments require an enormous amount of time to change. Even making punctual adjustments in a person's schedule, if we consider humanity, we will still feel sleepy when the sun goes down for thousands of years.

Similarly, evolutionary systems related to altruism still serve as the basis for human behaviour, despite notable differences in our social, psychological, and behavioural patterns from those experienced thousands of years ago. The evolutionary system continues to exert its influence. Such as, for example, in reciprocity, when interactions between individuals are repeated over time. If someone acts altruistically towards another person and that action is reciprocated in the future, both of you can benefit from the interactions throughout life.

This reciprocal relationship can evolve genetically, as genes that favour cooperation can be selected. In certain contexts, genes that favour cooperation provide adaptive advantages.

Thus, we can reflect on how genetic interactions and evolutionary patterns have shaped and continue to shape our societies and behaviours, even in the face of the radical changes that have occurred over the past few millennia.

The time has come when I will reduce the references to the past and thousands of years, as well as to the epigenetic question. However, make no mistake, these themes will always be present in my mind, and I hope you will also consider them.

When introducing the psychology of this topic, it's crucial not to limit our focus to our cultural understanding, both mine and yours. To do this, we need to turn to modern anthropology. Although we will later discuss what can be considered pathological and what should be healthy, it is essential to remember that this is about human behaviour, not restricted to our vision or culture, but rather a truly global approach.

First, we must consider everything we have discussed earlier, always remembering that in human psychology, the environment is an intrinsic part of what shapes behaviour. We must adopt a scientific perspective that is open to discussion, interpretation, and change, as society evolves, and concepts change.

One of the tools for analysing the basis of this fact in modern psychology is Game Theory.

Here I will propose a version adapted for this context of game theory, the reader will be able to read about the original Game Theory at any time, there is a vast literature on it, and it is within everyone's reach.

Let's imagine that there are two groups of prisoners, each of whom has a leader who represents them.

Both groups are sentenced to death.

However, they have been given the following opportunity to continue living if they face each other in a game of cooperation or betrayal. The rules are as follows:

The group that has a very low score does not change anything, there is no reward, and the death penalty will remain in force, that is, it is not a question of losing, but of scoring very low.

The group that gets a very high score can be given the freedom.

If the scores are similar and high, they won't win their freedom, but they can escape the death penalty. The results will be decided in the best of three rounds. They can choose to cooperate with the opposing team or cheat in each round. The results of rewards or punishments are adjusted according to the choices made by the two groups of players. If both cooperate, both receive a moderate benefit. If they both cheat, they both receive a very low score due to the negative consequences of their actions.

If one cooperates and the other betrays, the cooperator receives a very low reward, and the traitor receives a very high score.

The crucial feature of this Dilemma is that each player's choices affect future interactions. This allows more complex strategies, such as reciprocity, to come into play.

One notable strategy is the so-called "Tit-for-Tat," which starts by cooperating and then does what the opponent did in the previous round. This strategy encourages cooperation, as it rewards cooperative actions with cooperation and responds to betrayal with betrayal.

How does game theory relate to altruism?

Let's simplify this. Game theory is a way of understanding how people make decisions in varied situations that require strategic actions, such as games or social interactions.

In game theory, my intention was not so much about the theory itself, but more about how we apply strategies to gain advantages in different life situations. We look at the pros and cons of ourselves and the people around us, i.e. society. Sometimes, choosing to cooperate, even if it comes at a personal cost, can be considered altruistic. If people value everyone's well-being and choose to cooperate to have fairer outcomes, that's altruistic behaviour.

And in repeated games, people may alternate between selfish and altruistic behaviours, depending on the situation and what happened before. In some cases, players realize that cooperating in the long run is better for everyone,

then they choose to act altruistically to maximize overall well-being. To introduce this part of the psychology of altruism, I ask the following question: Do you still agree with the following information commonly found on the internet about altruism?

"An attitude that aims at the well-being of others, without taking into account particular interests." [Philosophy]

Auguste Comte (1798-1857). "A natural tendency of the individual who cares about others and, although it is spontaneous, needs to be perfected through positivist education, avoiding instincts related to selfishness."

Or what do you think of this dictionary definition?

[Brazilian Dictionary of the Portuguese Language] Spontaneous love for others; selflessness, philanthropy, proximity.

Chapter 3

Anthropology of altruism

The anthropology of altruism is a field of study that examines altruistic behaviour in different human cultures and societies. Altruism refers to the practice of acting for the benefit of others, often without expecting anything in return. Understanding altruism from an anthropological perspective involves analysing the cultural practices, social norms, and values that encourage or discourage altruistic behaviours.

So far, I have tried to demonstrate that altruism has its biological and evolutionary underpinnings, exploring how altruistic behaviour can be shaped by selective pressures over time.

The theory of biological altruism explores how altruistic behaviours can be adaptive, even when seemingly at odds with the logic of natural selection.

However, this behaviour depends on variations that can change significantly between cultures. What I consider an altruistic act in one society may not be seen in the same way in another.

Unlike in the United States, the Japanese prefer to help people work or study, in general, to achieve better living conditions, than to donate amounts of money to institutions that deal with social problems.

Altruism can also depend on social norms that play a key role in promoting or limiting it.

In some cultures, altruistic behaviour may be strongly encouraged and rewarded, while in others it may not be as emphasized. In societies that value individual autonomy, excessive altruistic behaviour can be viewed with suspicion. The Confucian tradition in China emphasizes the importance of social relationships, family, and respect for the elderly. These values contribute to an emphasis on collectively.

Countries such as Germany, France, the United Kingdom, and the United States are often considered examples of individualistic cultures, these countries have a strong emphasis on personal autonomy, individual fulfilment, and personal success. The "American Dream," for example. It often highlights the individual's pursuit of happiness and success.

There are also systems of reciprocity that encourage cooperation and altruistic behaviour in diverse cultures. Reciprocity can take different forms, such as direct exchanges of favours or the expectation that good deeds will be reciprocated in some way in the future. Religious and philosophical beliefs play an important role in shaping values related to altruism.

Many religious traditions, for example, emphasize the importance of helping others as part of a broader moral duty. When we are faced with issues or situations in which altruistic attitudes are the key part of a certain behaviour, in a certain context, or part of a certain society; It is up to us to understand what are the nuances that differentiate these cultures; and consequently, the motivations and structures of these thoughts.

By trying to apply universal concepts of altruism without considering these cultural nuances, we run the risk of acting inappropriately or even harmfully. An action that is perceived as benevolent in one culture may be interpreted completely differently in another. This can lead to misunderstandings, and resentment, or even contribute to mental health problems in individuals who feel misunderstood or unsupported in their specific cultural contexts.

Additionally, the imposition of external values without a deep understanding of the local culture can result in unsuccessful interventions. For example, psychological help strategies or social assistance programs can be ineffective or even counterproductive if they do not consider cultural particularities.

The Importance of the Worldview

Worldview is a concept that refers to how a culture interprets and understands the world around it;

influenced by its beliefs, values, traditions, and historical experiences.

Each society develops a unique worldview, shaped by its history and cultural context. The altruistic perspective in the worldview stands out as a fascinating example of reflecting the different approaches that different cultures may have towards altruism.

I propose two scenarios:

Star Worldview Tribe Scenario

In the worldview of the Stele Tribe, ancestral beliefs play a crucial role in shaping the collective vision. Teachings passed down through generations highlight collaboration as a principle.

essential for cosmic harmony. They believe that the stars represent the union of efforts, because, according to tradition, all the stars work together to assist the moon in its nocturnal birth. This celestial metaphor strengthens the idea that cooperation among tribesmen is vital to overcoming challenges and ensuring prosperity for all.

Contrast: Worldview of the Animalistic Tribe

In the Animalist Society, ancestral beliefs also play a significant role, but in a distinct way. Animalist society holds that altruism is a sign of strength and leadership, stemming from observing the strongest and most respected animals in their mythology. According to their traditions, the most admired leaders were those who showed generosity and care for their tribesmen. In this worldview, altruism is not a sign of weakness, but rather a lofty expression of strength and wisdom, capable of uniting and strengthening the community.

These ancestral beliefs profoundly shape each society's worldview, influencing its attitudes toward altruism and collaboration.

Star Tribe is based on collective vision, while Animalistic Society is based on leadership and individual strength, and particularly Altruism is a trait of nobility and leadership. Therefore, it is crucial to make sensible and culturally informed when dealing with altruistic issues.

This involves a willingness to learn about the cultural specifics of a community, listen to local voices, and tailor support approaches accordingly.

By doing so, we can ensure that our altruistic actions are truly beneficial and respectful in diverse cultural contexts, thereby promoting well-being and mental health more effectively and sustainably.

Chapter 4

Psychopathologies of the Altruism

It's worth noting that when it comes to the behaviour we've discussed below, we don't have a correct name to define certain behaviours. We still need more information to be sure of the definitions. But to make it easier, I'm using terms that psychologists often use to make diagnoses.

It is important to say that diagnosis cannot ignore the foundations of anthropology and biology. However, over time, things have become more complicated, and altruism has gained new characteristics, for example, in the matter of Narcissism.

In the intricate fabric of human relationships, a vast and complex field emerges, where the threads of altruism and narcissism intertwine, creating intriguing and paradoxical patterns. In this book, I'm trying to show that helping others and being selfish are like two sides of the same coin. It's like when you flip a coin – or heads or tails.

Here's an idea: if the coin has no very defined pattern, like, both sides have very similar heads and tails, can we still call it a coin?

Have you ever heard of the story "The Strange Case of Dr. Jekyll and Mr. Hyde" written by Robert Louis Stevenson? It's about a doctor who has a good side and a bad side, and the story explores how these two parts coexist in the same person. This story is so famous that we still use the phrase "Jekyll and Hyde" to describe someone who acts completely differently in different situations.

Altruism and narcissism dance in a delicate balance, each representing extremes that, if not kept in check, can lead to a complex web of distorted relationships. The challenge lies in recognizing the lures that attract us, questioning the motivations behind our actions, and cultivating a genuine understanding of what it means to be truly human.

Now, I'm going to clear up some distortions of what we understand about altruism in psychology. Don't worry, later, I'll talk about the truly healthy altruism that can be learned and practised. However, before that, let's go through some turbulence. So, be prepared. And remember the biological issues we've already discussed.

Nepotistic altruism: In the context of nepotism, there is a manifestation of family favours that can often distort meritocratic principles. One of the most serious examples of this pseudo-altruism is "La Cosa Nostra", which means "Our Things". This family society is considered the greatest expression of the original mafia based on the family clans of the Italian region of Sicily.

Governed by the famous "omertà", a code of silence to show total loyalty. The bait here is blood ties, the genetic connection that sometimes obscures fairness and justice, creating a breeding ground for vicious reciprocity.

Generally, some families have a reward process for members who join a common purpose, for example, inheritances where the beneficiaries are those who are in some way closest to the retainers of the resources. There is also the issue of benefits to children and close relatives who can maintain family traditions by studying the family career, as in a family where everyone for generations has been a doctor or even a military man. These new family members who follow this tradition will enjoy benefits, while others considered "black sheep" will live off their resources.

Here, too, the two-edged sword cuts, both on the side of the one who provides the resource, maintaining high advantages to those who succumb to family wills, and to those who know the benefits and put to this cause, if they mould the advantages offered.

Group altruism:

In the context of groups, compulsiveness can turn altruism into a paradoxical phenomenon. The unbridled pursuit of group approval can lead to

seemingly altruistic attitudes, but whose motivations are rooted in the desperate search for acceptance. The bait is social validation, and the price is the loss of authenticity. It often occurs in religious groups or societies, and social clubs that have philanthropic veins.

The dispute to provide for the needs of aid projects, and the integrity of churches or religious institutions that lack a physical continuity of these institutions or organizations, thus puts philanthropists in the spotlight, and from there a step in organizational political manipulation. In this way, altruism can be encouraged and practised as a manoeuvring machine, also establishing the duality of interests.

Personal Scope – Cognitive Distortions:

On a personal level, we have several types of altruism that, far from being common benefits, are ways of practising behavioural distortions.

Excessive altruism, when taken to the extreme, can result in several cognitive distortions. Like, for example, the thought of "Duty": "It is my responsibility to help everyone all the time, otherwise I will be considered an inefficient person."

Cognitive dysfunction is caused by the imposition of rigid and inflexible personal rules about altruistic behaviour, which results in an intense emotional burden and a constant sense of responsibility.

Pathological Altruisms:

The term "pathological altruism" refers to an excessively altruistic behaviour that may arise because of certain underlying conditions or diseases, i.e., the disease will give characteristics to behaviour. Altruism is the willingness to help or benefit other people, often without expecting anything in return.

However, when this behaviour becomes excessive to the point of harming one's health or well-being, it can be considered pathological.

There are several underlying conditions, which can contribute to pathological altruism. For example, some people may feel a compulsive need to help others due to anxiety disorders, obsessive-compulsive disorders (OCD), or personality disorders. They can find it difficult to say "no" to the requests of others, even when it hurts them.

Another example is the crippling altruism that arises in individuals with **eating disorders**, in which it is common for them to devote themselves excessively to ensuring that others are satisfied, neglecting their own needs. These behaviours may be motivated by a desperate search for validation or by an attempt to control your environment self through excessive care for others.

Emotional and sexual altruism

In the case of sugar daddies, a narcissistic pathological variant, financial favours often are exacerbated. to serve as a trap for emotional and sexual reciprocity the narcissistic element utilizes financial resources to attract the attention and dependence of others. In this game of co-dependency, the bait is material comfort, but the complexity of the relationships is revealed as a subtle dance between giving and receiving, where dependence turns into an emotional trap.

This form of narcissism disguised as altruism is slow, sometimes well-intentioned, not least because it has a pseudo-paternalistic element, but progressively escalates to the pathological relationship is learned behaviour or fit to be learned.

This process does not always develop actively on the part of the narcissist. Nowadays we don't know.

Well as the day of the hunt or the hunter, people who actively live in emotional and financial dependence or co-dependence are usually found.

Orbiting the narcissistic planets, waiting for the opportunity to access this magnificent attraction of benefit exchange.

As this process is often slow and nuanced in a game that "stretches and pushes" the exchange of benefits may seem altruistic, but it is not.

Altruism Motivated by Mind Reading

This is another unhealthy way for some people to have cognitively distorted altruistic practices. "Even though

they don't ask for my help, they need my assistance, even if they don't ask for it.

Mind-reading is characterized by the presumption of knowing the needs of others without direct communication, which can result in excessive and intrusive behaviours.

Overgeneralization:
It is another inappropriate characteristic. "If I don't help

or intervene constantly, no one else will, and it will all end in disaster."

This distortion entails drawing broad conclusions from isolated events, leading to an extremist view of one's altruistic responsibilities.

self-acceptance:

This next example is quite common in people with major self-acceptance problems, yet altruistic in its conception. "If I don't do everything for others, I'll be a selfish and worthless person."

Labelling oneself negatively based on specific behaviours feeds into the idea that personal worth is intrinsically linked to one's life ability to help others by creating intense emotional pressure. Disregarding one's own needs for the sake of others can lead to cognitive distortion that devalues self-care, resulting in exhaustion and resentment.

"My needs don't matter; The important thing is to help others." This attitude is always accompanied by exaggerated guilt and a bit mixed with masochism. "If

I take time for myself, I'm being selfish. "Feeling guilty about taking care of themselves can make a person feel bad about not taking care of themselves, which takes a toll on mental health.

Defensive Altruism:

Defensive altruism, which seeks to mask personal frailties; is a seemingly altruistic behaviour, but motivated, at least in part, by a need to protect oneself or mask personal frailties. In other words,

A person may act altruistically not only because they care about others, but also as a strategy for dealing with personal insecurities, gaining social acceptance, or building a positive image of themselves.

Thus, the paradox is again established, that caring about others is nothing more than the

compensation for personal emptiness, and narcissistic behaviours.

Evil altruism:

Motivated by dark intentions; is generally used as a strategy of political manoeuvre in wars, for example, offering exile and peace conditions by extraditing children from the enemy side to re-educate them in propaganda, or just to serve as material for abuse.

Masochistic altruism:

Masochism is liking to suffer or acquiescing to suffering as part of one's existence, also known as "martyrdom" or "martyr's gift", It describes an approach to altruism in which a person accepts and even pursues self-suffering for the sake of others. This pattern of behaviour can result from a variety of motivations and beliefs, but it usually involves a willingness to sacrifice one's well-being for the benefit of others.

Some characteristics associated with masochistic altruism include:

Extreme Personal Sacrifice:

Many masochistic people in altruism sacrifice their interests, comfort, and happiness for others. This can manifest itself in situations where the person is struggling or hampered to meet the perceived needs of others.

Sense of Duty and Obligation:

The person who takes this approach often feels a strong sense of duty and obligation towards others. This can lead to a sense that it is your role to suffer in the name of the well-being of others.

Refuse to accept help:

People with masochistic tendencies may find it difficult to accept help from others, as they view the act of receiving assistance as a violation of their commitment to personal sacrifice.

Self-Worth through Suffering:

There is a possible self-worth associated with suffering. A person's worth can be directly linked to the amount of sacrifice they make for others, which can lead to a constant search for situations of suffering.

Exceptions of the Masochist model:

As a writer, it's hard to ignore the desire to dodge the theme of sacrifice in war—to die for one's colleagues, for one's country, or in the name of an ideology. Certainly, this represents sacrificial altruism, but even if for the sake of science, we must consider all possibilities, most of the time, we cannot label it as masochistic.

First, this is a matter of duty; most of the time it has nothing to do with desire. There are strong influence of kinship altruism and subsequently, when in confrontational situations, this feeling is linked to specific situations where rationality and emotions are taken to extremes, such as in bloody battles and unimaginable scenarios for those who are not present in that local moment in time.

These situations should be exempt from judgment, at most, we can look at the group issue, but still, I think this is a separate issue in this context. Medals of honour should not be regarded as selfless acts subject to judgment. No one has the elements to interpret the facts and question the altruism of the heroes, no one has that kind of authority to do so.

The Possible Hidden Resentment:

Although they exhibit a willingness to sacrifice, some people who engage in this behaviour may feel angry toward others for not recognizing or valuing their sacrificial effort appropriately. Masochistic altruism can have negative implications for the emotional and physical well-being of the person involved. It's important to note that healthy altruism involves finding a balance between helping others and taking care of yourself. Constant personal suffering can lead to exhaustion, burnout, and negative impacts on a person's quality of life.

Metacognitive dissonance:

It refers to the awareness and understanding of one's own thought processes. It involves thinking about thinking, such as reflecting on your own cognitive abilities, monitoring your thought patterns, and making judgments about your mental processes. Where the giver deludes himself about his true intentions; Metacognitive dissonance can create an internal conflict, leading the giver to distort their perceptions to avoid the discomfort associated with the inconsistency between their actual intentions and their conscious image. It is important to note that these processes can be complex and are often not completely clear, therefore, there is a dissonance. It brings the inception of:

Cognitive dissonance:

Cognitive dissonance within metacognition occurs when an individual becomes aware of a misalignment between their held beliefs or perceptions about their thinking abilities and the actual cognitive processes they observe. This incongruence can lead to a sense

of discomfort or tension, prompting the individual to adjust constantly their beliefs, creating a pattern of inconsistencies. Missing the point of promoting a balanced approach to altruism, one that considers both the needs of others and one's own is essential for maintaining a healthy relationship with oneself and others.

Social Validation:

Imagine a person who donates generous amounts to charity and consciously believes that they are doing so solely to help those in need. However, at the metacognitive level, there may be a disconnect, and that person may not realize that they are seeking social validation. This desire can manifest itself when she makes a point of publicizing her donations on social media to receive praise and recognition.

Self-Deception and Rationalizations:

A regular volunteer at an animal shelter may genuinely believe they are there just to help the homeless animals. However, at the metacognitive level, this person may be using self-deception mechanisms. For example, she may justify her actions by telling herself that she is compensating for a past negative action, while unconsciously seeking to feel better about herself.

Selfish Motivations or Protagonism Seekers:

It starts by ignoring signs of selfish motivations. An individual who organizes charitable events may be convinced that their sole motivation is to help the

community. However, at the metacognitive level, he may ignore signs of selfish motivations, such as the personal pleasure he derives from being recognized as an exemplary benefactor. This person may repress these thoughts to maintain an altruistic image of themselves exchanging it for a positive or protagonist self-image.

Consider someone who devotes considerable time to volunteer activities, such as mentoring young people in difficult situations. Consciously, this person may believe that their only motivation is to help others. However, at the metacognitive level, she may resist recognizing any sign of selfish motives, such as the unconscious impulse to consider herself morally, culturally, etc. superior to others, which could threaten their positive self-image.

Protective Altruism:

Protective altruism refers to overzealous and controlling behaviour where someone tries to help or protect another person in a way that may be stifling or harmful. This form of altruism often arises from the good intention to care and protect, but it can become harmful when it crosses healthy boundaries.

Imagine someone who is always mindful of the actions of a friend or family member, interfering in every aspect of their lives, even when it's not necessary. This behaviour may arise from genuine concern, but overprotectiveness can stifle the other person's freedom and autonomy, hindering their ability to grow, learn from their mistakes, and develop coping skills.

Protective altruism has no place, it occurs without a legitimate need. The person believes that he should do something or give his opinion even if there is no need, that is, requested. This can be motivated by several things, such as anxiety, over-protagonism, fear of bad things with loved ones, or the need to control. Sometimes, those who act in this way may believe that they are doing what is best for the other person, but it is important to recognize when the protection becomes excessive and harmful.

Pseudo-altruism:

Attitudes whose altruistic actions are motivated by selfish interests.

False altruism is when people act altruistically for their selfish interests. In other words, the person may appear to be acting for the benefit of others, but they are seeking to meet their own needs or goals.

There are several reasons why someone might engage in false attitudes of solidarity. This can include a desire to be socially recognized, receive favours, build a positive image of oneself, or even manipulate people to achieve one's personal goals. The person may act in a seemingly altruistic manner, but in fact, they are more focused on getting benefits for themselves than genuinely helping others. These actions can be deceptive, since on the surface they may seem beneficial, but the real motivation is selfish.

It is important to distinguish between true altruism, which involves a genuine concern for the well-being of others, and pseudo-altruism, which is more oriented toward meeting one's own needs under the guise of helping.

Psychotic altruism:

An individual with psychotic altruism may have the sincere conviction that they are acting beneficially, but their disorganized perceptions can distort their understanding of what is helpful to others. These actions of solidarity may be motivated by irrational beliefs or delusions related to the need to protect, or save others from, imaginary dangers. For example, someone who is in a psychotic state may imagine that they are doing actions to protect people from threats that do not exist, which results in behaviours that, depending on the shared reality, seem strange or inappropriate.

Paranoid altruism:

The term "paranoid altruism" designates a situation in which excessive care for others becomes an unbearable burden for the individual who practices it. In this situation, empathy, guilt, and depression can emerge as unwanted companions.

People feel obligated to care for others, but this can be an emotional burden, which can cause them to get lost in the pursuit of other people's well-being.

They always feel persecuted by the burden of feeling obligated to do their altruistic acts. Distorted

empathy, being the ability to understand and feel the emotions of others, can lead to an over-absorption of those emotions, contributing to emotional overload.

Guilt can arise when the person feels that they are not doing enough to help others, or it can result in a constant worry that their actions do not meet the expectations of others.

Depression can happen because they are tired and tired of caring for others, even if it costs their own life. When they care a lot, they attract people who benefit or benefit from their kindness, which makes them feel heavier emotionally and makes life more difficult.

Chapter 5

Professional Dependents

They say that where there's smoke, there's fire. That's true, but often, where we see smoke, the fire has already done its damage or is about to start. In any case, this expression persists because it carries truths.

Talking about altruism is tricky without considering the stark reality of addiction. Using the previous analogy, where there is altruism, there is dependence. It can be starting, in progress, or newly established.

As mentioned before, there is a certain symbiosis encouraged and promoted by the altruist himself. In the same way, on the other hand, there is a professional dependence when the relationship is symbiotic, where one life helps the other to subsist.

We might even use the term "co-dependency," because that's what happens – one depending on the other to play their roles in life.

The professional addict is experienced and understands the psychology of the altruist. I call him a professional because he has had previous experiences, and several beneficial relationships with altruists, which have provided them, as dependent professionals, with the necessary experience to continue exploring the benefits of this lifestyle.

This relationship can be financial or not, but it is related to obtaining advantages from those who have more resources to get what they need. The professional dependent understands that the altruist needs a reason and that one must submit to the altruist's operational methods.

The professional addict knows how to manipulate time to his advantage, almost as if he were talking about the theory of relativity. He also knows how to manipulate space, i.e. proximity and infinite distance.

It's not just about exploiting wealth, as professional dependents are more interested in resources. They know that wealth cannot be transferred to them, but resources are always available. These resources are not just money, they include emotional, structural, managerial, psychological, and support. However, material resources are essential.

Where there is an altruist, where the situation of an altruist is known, we will certainly find professional dependents.

Being a professional addict implies giving up certain things, such as having one's own opinions and unique lifestyle. For the professional dependent, it is necessary to be aligned with the altruistic vision, embracing the donor's perspective as their own. However, this does not mean being devoid of individuality; On the contrary, addicts do contribute in some way to the well-being of the altruist. The goose that lays the golden eggs should be kept forever, very happy and healthy.

Addicts are not simply passive receivers; They offer something valuable, which can range from physical pleasure to attention and submission, identifying the specific needs of the altruist.

There's a subtle ability to offer what appeals to these altruistic personalities, often exploiting their weaknesses. We can call them true experts in a kind of "emotional fishing".

Toxic addiction is a complex phenomenon that deserves further analysis, it becomes apparent that relationships marked by toxic addictions are not just simple connections, but rather intricate traps. What may initially seem like mutual support turns into an intricate emotional tangle.

Addicts, endowed with manipulative skills, can establish emotional bonds so strong that they make it difficult for the altruist to break free. This emotional connection causes the altruist to get stuck in a cycle of demands and expectations that overwhelm the addict.

It is crucial to understand that, in this scenario, the nature of addiction goes beyond the simple emotional connection and turns into a complex network of obligations, commitments, and psychological pressures.

The altruist often finds himself entangled in this entanglement, experiencing a sense of powerlessness in the face of the growing demands of the addict. Thus, toxic addiction is a complicated emotional trap, and it takes care and awareness to unravel and break the bonds that cause it.

Chapter 6

True Altruism

I believe that the main quality of true altruism is empathy, which means understanding the needs of others and understanding situations and contexts. Being empathetic involves getting to know these needs deeply, understanding the reasons behind them, and seeing the details, as if we were putting ourselves in the shoes of those who need help.

True altruists are like analysts, never acting in isolation. Even if they make decisions on their own, they always seek a second opinion, forming a kind of team, whether formal or not. These people are also like architects, carefully designing the cover and protection they will offer, considering proper proportions and measures to be effective when helping repeatedly.

Another important virtue is humility. True altruists don't consider themselves superior to anyone, they understand that each situation is unique, and they don't see themselves as saviours. They act to compensate and balance, without overstepping.

boundaries or disrespect, keeping respect as the main motto. Moreover, true altruism does not seek propaganda.

As the biblical phrase suggests, they give without expecting recognition, acting with the pure intention of helping. "But when you give to the needy, do not let your left hand know what your right hand is doing, NIV Mathew 6:3". Not only does this prevent them from relying on praise, but it also shows that their focus is on empathy for the helping process, not on seeking external recognition.

Altruism has a process that follows three important steps: the beginning, the middle, and the end.

Beginning: It all starts with empathetically understanding the other's needs and devising a plan to help.

Middle: The intermediate stage is the execution of the plan, aimed at promoting mutual well-being. It is crucial to act in a balanced manner, ensuring that those who offer help do not commit themselves beyond their means and that those who receive it benefit effectively, without the need for frequent interventions.

End: The success of altruistic action is marked when the need of the person being helped is satisfactorily fulfilled, without causing harm or excess. This represents the end of the intervention, indicating that altruism was effective. In summary, altruism involves understanding, acting balanced, and achieving success by meeting the needs of those receiving help.

"Of course, I could address several aspects of altruism, highlighting its importance and healthy ways to practice it. However, I prefer to conclude with a simple concept crucial to assessing the health of this process.

The altruist must be the target of his altruism, learning to receive and always being able to understand the act of receiving.

In aviation, flight attendants teach a vital rule about flight safety: In the event of cabin depressurization, oxygen masks will fall from the ceiling. The instruction is to put the mask on yourself first and then on the child or dependent person. If you put the mask on the child first, you run the risk of losing consciousness before ensuring your safety, leaving the dependent person alone and awake.

Learning to receive help begins when we know how to help ourselves. We are only able to help others when we are fully aware. In addition, receiving help from others strengthens the feeling of healthy co-dependency. Understanding that we don't have everything and that material possessions don't indicate insufficiency or superiority is key because as human beings, we all need it constantly. We need to understand this dynamic. I recognize that there are other nuances and more material on this fascinating topic, but so far, that's all I wanted to share.

Chapter 7

Altruistic Attitudes and the Tool of Prosperity

In this last chapter, I will address altruistic attitudes. Now that I'm going to talk about behaviour, I'm going to rescue what I've explored throughout this book at various points, especially at the beginning. I called the reader's attention to the question, "Do you remember that time?" How many things have changed along the way?

The cars of the past are no longer like those of today, but they are still cars. The clothes of the past, although we don't want to wear them again because they are out of fashion, are still clothes. Today we live in the behavioural biology and anthropology of the past, which still form the basis of what exists. Behaviour continues to be and always will be guided by the anthropological, biological, and psychological bases of us human beings in the most fundamental sense of our existence.

When I brought examples of psychopathologies that can interact with altruistic attitudes, I was already speaking of modern changes and expressing that some attitudes have evolved over 356,000 years,

becoming more complex in the formation of the mind and behaviour. Society has also become extremely more complex, influencing behaviour, anthropology, epigenetics, etc.

However, it is important to remember that development does not neglect the foundations, just as a house is not just the foundation; There is no house without a foundation. You must build things on top of it and use layers to build something. When we look at something that has gone through evolution, through development, we are seeing the final making of something that began with a foundation, and the foundation is still there, sustaining all things.

In today's world, managing resources is critical. 80% of the world's resources are distributed among only 10% of the population, which is the rich class.

The middle class gets 15% of the resources.

However, most of the world's population, those who are below the median line, live on only 5% of the planet's resources, and they are 75% of the population.

While these statistics lead to deeper discussion and reflection, in the context of altruism, it is crucial to consider balance. You can't be altruistic without thriving. As I mentioned previously, being altruistic when you don't have resources is pathological, leading to illness, loss of altruistic qualities, and turning action into a frustrated attempt to help, generating depression. The tenor and quality of this desire are indisputable; certainly, it is an altruism embryonic. However, in this case, pathological, unhealthy altruistic behaviour can be established.

Therefore, altruism needs an effective tool, a modern tool. Prosperity functions as a process of curbing impoverishment; Wanting to prosper is the point of return and a path of time and effort, provided with goals.

Prosperity here is not just financial prosperity, but it is prosperity in a more global, broader sense of our existence, which relies on a multitude of resources.

Improving life in many ways, from knowledge to specialization, from the spiritual to the material, where knowing and being able to teach is a great resource., Educating with conditions is another, improving health is another, etc., how a life can prosper is a sea of possibilities.

It is the path that the naledi men had to take to deposit their loved ones in the urns inside the cave depended on the basic resources, for this, they had to be able to get to the place of burial, wood, knowledge for the fire, food for the path, proportional resources to complete their journey, had to have the ability to generate resources for this purpose. Yet the things that drive altruism remain the biological and anthropological underpinnings mentioned here. The tools have changed, the contexts have changed, and they continue to change. when we mention altruistic attitudes, we are not referring to large groups that invest large amounts in some project, nor to benefactors who identify a situation and want to contribute to solving it. I'm not talking about people who make large donations, as we've discussed in previous chapters. I'm referring to us, all of us who are in some way connected to other people.

We observe situations in which many people suffer before us. These people may be poor, face financial or material needs, or they may also have spiritual, emotional, health needs, etc.

The altruistic attitude is when there is someone next to them, or someone watching in some way, applying the force of altruism to them. A force that arises from within us when we believe that we can help someone out of a difficult situation or improve their condition by using the various extra resources we have.

By no means is an altruistic attitude free from the pathologies or psychopathologies that we have already discussed. It is not necessary to be wealthy or possess extravagant resources to represent a trait that promotes better health for those who need help. It's a misconception that having more resources puts us in an advantageous position. I mean if you have resources and you employ the altruistic attitude underneath a pathological behavioural condition, you're not doing altruism and it's not beneficial to either side.

Another important consideration is that reaching out and helping should be a natural, empathetic conscious, and elaborate act within our possibilities, without harming our own lives. That means giving what's left over; If there is no leftover, we are not altruistic. If we take from ourselves, we are not acting altruistically. If we want to be truly altruistic, we must give away what we have in excess, whether it be time, clothing, food, attention, money, companionship, or love. The key is to have more than enough because in giving, we offer the surplus, not what we need.

Also, you can't balance the pain with your pain. This is more than stupid; It's impossible. If a person has 100% pain and you donate 50% of your health to them, then now they are.

Two people in need of 100% help, and no, the math isn't wrong, think about it. It's crucial to understand that helping someone doesn't mean compromising your ability to live.

Altruistic attitudes should be encouraged in our groups, families, work environments, and social life. They need to be something intrinsic to our environment.

Unfortunately, these days, the mentality of "where you can eat three, you eat four; where they can eat four, they eat five" is already out of fashion; That's not how the world is thinking anymore. The truth is that the prevailing idea is: "Where you can eat three, you can eat only one" and he lives better. This idea goes against human nature. When we neglect altruism in our thoughts and lives, we go against our human nature, making us inferior to the beings who lived thousands of years ago. To recover altruism is to rescue the perspective of the other human being, it is to look beyond ourselves. We must stop ignoring the needs we perceive around us. Looking at someone next to us, or with whom we have a relationship, means ceasing to be narcissistic.

It means to stop seeking personal advantage all the time. We must abandon the self-centred gaze, the armoured heart, and the hard hand. We must offer solutions. Real and committed, within our possibilities, to change situations.

When getting involved in a cause of others or a pain that is not ours, doing so with respect and silence, respecting the fragility and dignity of the other is fundamental. We should not draw attention to ourselves when we help, avoiding superficial and uncompromised solutions. It is necessary to reflect on how many, and how much is needed in all perspectives before acting.

We received help many times and how rewarding it was to be able to help other people. If you believe that you have never needed help from anyone, then you are passive and the target of an altruistic attitude. Your perception of life is devoid of gratitude and understanding of others. Start practising altruism. It's not complicated. It doesn't take much, just a surplus of what you already have. And this surplus can be willingness, time, listening, sharing, or even financial resources. Volunteer, offer your skills, and make your resources available together with others. Donate, be charitable, and help raise funds. Often, we can't provide the resources directly, but we can make them viable, enabling is helping the path flow, it may be that you know about logistics or management. Offer emotional support, and mentorships are a huge resource, as your experience can be what changes the perspective of someone or groups because many people just need it in difficult times. Add that "extra" when someone needs it, but don't help those who don't. Be sensitive to understanding the vulnerability of others. Help those who are trying to get back on their feet.

Promote awareness of true altruism in society. This will result in a society that is more elevated and prepared for positive change.

Be respectful, courteous, polite, and understanding of the dignity of the other. Kindness is a set of things, including education, thoughtfulness, and the fabulous practice of empathy. Be empathetic, look at the other through their eyes, and understand their suffering. Preserve the medium in that you live, nature, and the planet, because in doing so, you are being selfless to future generations.

Look at the past with a benevolent perspective. To rescue altruism is to rescue the essence of humanity in us, something transmitted through generations by our people's genetics. Don't allow nonsensical ideas to robotize you; or dehumanizing narratives, many of them are part of mass movements with purely financial and commercial interests. Before concluding, I would like to remind you once again of those. Humanity's ancestral cousins, the Naledi, who's main

activity in the formation of life was to value and give importance to others of their kind. All they possessed was an attitude driven by something internal to their being, and the resource was the place to give meaning to the lives of those who no longer lived, their altruism was to conserve the love of the ones of their kind, to walk in a dark and dangerous cave to give protection and eternal continuity to their own.

Lastly, prosper, seek to prosper in all that you have, and accomplish. It is the surplus in you that will make the difference in making an effective instrument for true altruism. Prosperity will enable you to help. Good way.

I wish you prosperity.

Bibliography:
Read more about the topic in the following references:

[1] Homo Naledi - Lee R Berger, John Hawks, Darryl J de Ruiter, Steven E Churchill, Peter Schmid, Lucas K Delezene, Tracy L Kivell, Heather M Garvin, Scott A Williams, Jeremy M DeSilva, Matthew M Skinner, Charles M Musiba, Noel Cameron, Trenton W Holliday, William Harcourt-Smith, Rebecca R Ackermann, Markus Bastir, Barry Bogin, Debra Bolter, Juliet Brophy, Zachary D Cofran, Kimberly A Congdon, Andrew S Deane, Mana Dembo, Michelle Drapeau, Marina C Elliott, Elen M Feuerriegel, Daniel Garcia-Martinez, David J Green, Alia Gurtov, Joel D Irish, Ashley Kruger, Myra F Laird, Damiano Marchi, Marc R Meyer, Shahed Nalla, Enquye W Negash, Caley M Orr, Davorka Radovcic, Lauren Schroeder, Jill E Scott, Zachary Throckmorton, Matthew W Tocheri, Caroline VanSickle, Christopher S Walker, Pianpian Wei, Bernhard Zipfel (2015) Homo naledi, a new species of the genus Homo from the Dinaledi Chamber, South Africa eLife 4:e09560 https://doi.org/

Archaeology of Altruism:

1. Mithen, S. (1996). The Prehistory of the Mind: The Cognitive Origins of Art, Religion and Science. Thames & Hudson.
2. Shennan, S. (2009). The Archaeology of Human Ancestry: Power, Sex and Tradition. Routledge.

Epigenetics of Altruism:

3. Carey, N. (2012). The Epigenetics Revolution: How Modern Biology is Rewriting Our Understanding of Genetics, Disease, and Inheritance. Columbia University Press.
4. Bruder, C., & Michels, K. B. (2013). The Epigenetics of Twins. Springer.

Anthropology of Altruism:

5. Flescher, A. M., & Worthen, D. L. (2007). The Altruistic Species: Scientific, Philosophical, and Religious Perspectives of Human Cooperation. Templeton Press.

6. Wrangham, R. (2009). Catching Fire: How Cooking Made Us Human. Basic Books.

Psychopathology of Altruism:

7. Stout, M. (2006). The Sociopath Next Door: The Ruthless Versus the Rest of Us. Broadway Books.
8. Cleckley, H. (1988). The Mask of Sanity. Emily S. Cleckley.

Nepotistic Altruism:

9. Dawkins, R. (2006). The Selfish Gene. Oxford University Press.
10. Wilson, E. O. (2012). The Social Conquest of Earth. Liveright.

Group Altruism:

11. Axelrod, R. (2006). The Evolution of Cooperation. Basic Books.
12. Haidt, J. (2012). The Righteous Mind: Why Good People are Divided by Politics and Religion. Pantheon.
13. Personal Scope – Cognitive Distortions:
14. Kahneman, D. (2011). Thinking, Fast and Slow. Farrar, Straus and Giroux.
15. McRaney, D. (2011). You Are Not So Smart. Gotham Books.

Pathological Altruism:

16. Oakley, B., Knafo, A., Madhavan, G., & Wilson, D. S. (Eds.). (2012). Pathological Altruism. Oxford University Press.

Emotional and Sexual Altruism:

17. Fisher, H. (1992). The Anatomy of Love: A Natural History of Mating, Marriage, and Why We Stray. Ballantine Books.
18. Chapman, G. (2015). The Five Love Languages. Northfield Publishing.

Altruism Motivated by Mind Reading:

19. Nichols, S. (2004). Mindreading: An Integrated Account of Pretence, Self-awareness, and Understanding Other Minds. Oxford University Press.

Defensive Altruism:

20. de Becker, G. (1998). The Gift of Fear: Survival Signals That Protect Us from Violence. Dell.
21. Masochistic Altruism:
22. Bancroft, L. (2002). Why Does He Do That? Inside the Minds of Angry and Controlling Men. Berkley.

The Possible Hidden Resentment:

23. Goleman, D. (1995). Emotional Intelligence: Why It Can Matter More Than IQ. Bantam.

Metacognitive Dissonance:

24. Tavris, C., & Aronson, E. (2007). Mistakes Were Made (But Not by Me): Why We Justify Foolish Beliefs, Bad Decisions, and Hurtful Acts. Mariner Books.
25. Protective Altruism:
26. Yalom, I. D. (2002). The Gift of Therapy: An Open Letter to a New Generation of Therapists and Their Patients. Perennial Harper.

Pseudo-Altruism:

27. Zimbardo, P. (2007). The Lucifer Effect: Understanding How Good People Turn Evil. Random Hous

Psychotic Altruism:

28. Hare, R. D. (1999). Without Conscience: The Disturbing World of the Psychopaths Among Us. Guilford Press.

Paranoid Altruism:

29. LaPorte, D. J. (2015). Paranoid: Exploring Suspicion from the Dubious to the Delusional. NYU Press.
30. Dependent on Altruism:
31. Beattie, M. (1987). Codependent No More: How to Stop Controlling Others and Start Caring for Yourself. Hazelden.